Bending Solid Wood To Form

U.S. Department of Agriculture
U. S. Forest Service

Fredonia Books
Amsterdam, The Netherlands

Bending Solid Wood to Form

by
Edward C. Peck

for U. S. Forest Service

ISBN: 1-4101-0905-4

Reprinted from the 1968 edition

Fredonia Books
Amsterdam, The Netherlands
http://www.fredoniabooks.com

Bending Solid Wood
To Form

CONTENTS

BENDING SOLID WOOD TO FORM

By EDWARD C. PECK,[1] *Technologist*
Forest Products Laboratory,[2] *Forest Service*

INTRODUCTION

Wood bending is an ancient craft that is of key importance in many industries today, especially in those that manufacture furniture, boats and ships, agricultural implements, tool handles, and sporting goods. Of the several methods commonly used to produce curved parts of wood, bending is the most economical of material, the most productive of members of high strength, and perhaps the cheapest.

Long experience has evolved practical bending techniques and skilled craftsmen to apply them. Yet commercial operations often sustain serious losses because of breakage during the bending operation or the fixing process that follows. There is a longfelt need for more reliable knowledge about: (1) Criteria for selection of bending stock; (2) better methods of seasoning and plasticizing wood for bending; (3) more efficient machines for the bending operation; (4) techniques for drying and fixing the bent part to the desired shape; and (5) the effect of bending on the strength properties of wood.

This handbook is based on results of research on wood bending and related information as developed at the United States Forest Products Laboratory and other laboratories over a period of years, and on investigations and observations in furniture factories, ship and boat yards, and other plants engaged in commercial wood bending. At the outset, however, it should be understood that much fundamental information is still lacking about basic factors and that the conclusions and recommendations given here are therefore limited.

WHAT HAPPENS WHEN WOOD IS BENT

It is common knowledge that a very thin strip of wood can be easily bent with the hands to quite sharp curvature. In making baskets and other products, thin veneers are bent by hand and held in place by weaving them together or attaching them to other parts.

---

[1] Acknowledgment is made to various members of the Forest Products Laboratory who have been consulted on subjects relating to their fields of investigation and from whom much helpful information has been obtained.
[2] Maintained at Madison, Wis., in cooperation with the University of Wisconsin.

Such bending is done without treating the wood. In bending thick pieces of solid wood, however, softening with steam or hot water or plasticizing with chemicals is essential.

When a piece of wood is bent, it is stretched, or in tension, along the outer (convex) side of the bend and compressed along the inner (concave) side. Its convex side is thus longer than its concave side. This distortion is accompanied by stresses that tend to bring the bent piece back to its original straightness. The purpose of softening wood with moisture and heat or plasticizing chemicals is to restrict the development of these stresses.

Plasticized wood can be compressed considerably but stretched very little. The objective in bending, therefore, is to compress the wood while restraining the stretching along the convex side. Although various devices are used to accomplish this, the most efficient yet found by the Laboratory is the tension strap, complete with end fittings such as end blocks or clamps and a reversed lever.

Figure 1 illustrates what happens when wood is bent. (See appendix p. 31 for a mathematical analysis of the bending action and reaction.) The resistance of wood to bending forces at the point of contact of the stick and the bending form, point O, is approximately constant at all times as bending proceeds. In effect, the unbent part of the stick is a lever and point O a fulcrum as bending force is applied. As this lever arm, represented by the distance $L + X$, becomes shorter, the bending load (P) needs to be increased. At the same time, however, the resistance to end pressure, represented by P', is also constant, and as a result, PX must remain constant.

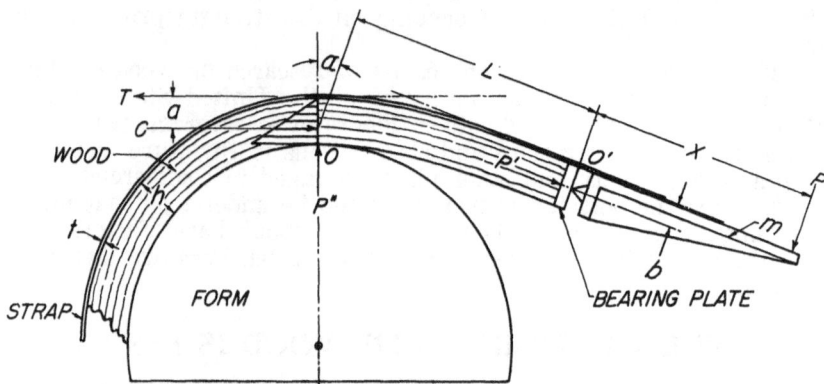

M—12386—F

FIGURE 1.—Diagrammatic sketch of the mechanics involved in wood bending under end pressure. (Device: A tension strap and an end block equipped with a pivoted bearing.)

Since X, representing the length of the end fitting, is a constant, the load P must also remain constant. However, P cannot remain constant, because as the stick is bent, the lever arm $L + X$ becomes shorter. Therefore, P must be increased as the bending progresses.

If P increases, PX increases, and the end pressure is thereby increased. Since the resistance to end pressure, P', remains constant, the increased end pressure must result in crushing of the end of the piece or in back bending or buckling. As a result, either too much end pressure is applied during the late stages of bending or too little during the early stages.

In spite of these shortcomings, apparatus of this type may produce successful bends, particularly bends with a small angle of curvature. A common fault of much commercial bending apparatus is in the lever arm that extends beyond the end block. If the lever arm to which the bending force is applied is relatively short, the moment that resists rotating or overturning of the end block may be too small. In such cases, it is impossible to keep the whole stick in compression even when bending to a small angle of curvature.

A simple device that automatically compensates for the higher end pressure induced by increased bending loads is a reversed lever (fig. .2). This device exerts pressure on the convex face of the stick, and so prevents the end block from overturning. Preventing the overturning of the end block assures the application of end pressure. Because the pressure against the block is generated by the stick itself as it is bent, the force required to make the bend has no effect on the amount of end pressure thus generated.

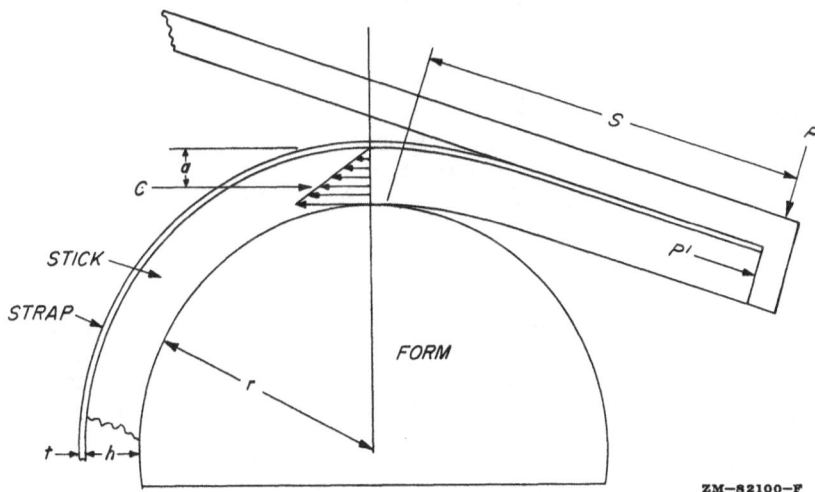

ZM—82100—F

FIGURE 2.—Automatic regulation of end pressure. (Device: A tension strap and an end block equipped with a reversed lever.)

The reversed lever must extend back as far as the last point of contact between the wood and the form in order to regulate end pressure and prevent reverse bending near the end of the stick. If the stick need not be bent throughout its length, the reversed lever will produce an almost uniform end pressure during the bending operation. With this device it is unnecessary to use a pivoted bearing. (See appendix, p. 33, for an approximate mathematical analysis of the stresses in the steel strap and in the wood.)

SELECTING BENDING STOCK

The selection of a species of wood to use in a manufactured product containing a bent member of slight to moderate curvature is largely governed by suitability and availability of the species for the product. If the curvature is to be severe, however, the species of wood must be selected primarily for its bending quality.

The bending quality of wood varies widely not only among the different species but also within the same species. Insufficient data are available, however, to permit the listing of species in the exact order of their bending qualities. As a rule, the bending quality of hardwoods is better than that of softwoods, and the latter are seldom used in bending operations. Yew and Alaska-cedar are exceptions; Douglas-fir, southern yellow pine, northern and Atlantic white-cedar, and redwood are often bent to moderate curvature for ship and boat planking.

Stevens and Turner[3] rated a few American species of known good bending quality in the following descending order: white oak, yellow birch, and white ash. Mahogany from Central America was rated as having moderately good bending quality and as superior to khaya, "African mahogany," in this respect. These ratings are based on the radius of curvature for steamed 1-inch material at which breakage does not exceed 5 percent when the wood is bent under end pressure.

Numerous hardwood species were ranked at the United States Forest Products Laboratory on the basis of percentage of breakage sustained during a uniform bending test made without end pressure. These species, in descending order of bending quality, were as follows: Hackberry, white oak, red oak, chestnut oak, magnolia, pecan, black walnut, hickory, beech, American elm, willow, birch, ash, sweetgum, soft maple, yellow-poplar, hard maple, chestnut, water tupelo, cottonwood, black tupelo, mahogany, American syca-more, buckeye, and basswood. The species commonly used in industry for making bent members are: White oak, red oak, elm, hickory, ash, beech, birch, maple, walnut, mahogany, and sweetgum.

An extensive study of wood cut from 20 white oak trees grown in four different localities in Ohio and Kentucky revealed that the bending quality of wood from the four localities varied. And, in most cases, the bending quality of the wood cut from trees in the same locality also varied. An attempt was made to correlate bending quality with specific gravity, rate of growth based on the number of growth rings per inch, end pressure developed during bending, and standard toughness values. No good correlation was found. In fact, the wood of highest bending quality was almost identical to that of lowest bending quality in specific gravity and rate of growth.

Despite the fact that little correlation has been found between basic physical properties and bending quality, stock can be selected with reasonable assurance that it will bend without undue breakage. The principal precaution is avoidance of stock that contains strength-

---

[3] STEVENS, W. G., AND TURNER, N. SOLID AND LAMINATED WOOD BENDING. 71 pp., illus. Forest Products Research Laboratory, Princes Risborough, England. 1948.

reducing defects. These defects are decay, cross grain, knots, shake, pith, surface checks, and brash wood. Even wood containing incipient decay fails under slight tensile stress and cannot be compressed nearly so much as normal wood (fig. 3).

M—80834—F

FIGURE 3.—A bent chair part of elm containing incipient decay. The impaired strength properties caused the piece to develop compressive failures in the form of abrupt wrinkles during bending.

Straight-grained wood is much less likely to fail during bending than cross-grained wood. The grain should slope not steeper than 1 inch in 15 along the length of the piece. Local cross grain is also to be guarded against; it is weak under bending loads.

Knots are objectionable because they are invariably accompanied by distorted grain and also because they resist compression.

Shakes, which are longitudinal separations parallel to the annual rings, are responsible for failures in shear during the bending operation. Pith introduces similar lines of weaknesses. Surface checks, which are cracks perpendicular to the annual rings, are likely to cause compressive failures if on the concave side of the piece. On the convex side, they are not so detrimental. Surface checks in combination with cross grain are likely to cause pyramid-shaped pieces to be forced out of corners that are in compression and slivers out of corners subjected to slight tension. Exceptionally lightweight wood is likely to be brash and fail during bending. Even if it is bent successfully, it may not have the necessary strength in service.

Minor defects in wood, such as small pitch streaks, burls, and small checks, are permissible if they are located beyond the part to be bent or on the convex side of either mild or severe bends. They may be allowed on the concave side of mild bends, but with severe bends such defects cause concentration of stresses and compressive failures are likely to result.

SEASONING BENDING STOCK

From the standpoint of the bending operation alone, most curved members can be produced from green stock. However, the moisture content of bending stock should also be suitable for the drying and fixing of the bend and for the drying process needed to bring about a moisture content suitable for the finished product. If it is not, the wood is likely to check, split, and shrink excessively. For these reasons, green or partly seasoned stock is not suitable for many bent items.

On the other hand, dry wood is not sufficiently plastic to bend well even when heated to a high temperature. Most kiln-dried lumber is not suited for bending unless it is steamed or boiled enough to absorb the moisture needed to make it sufficiently plastic to bend well. Soaking it in water before it is steamed or boiled is helpful.

To avoid both drying troubles with wood that is too green and bending difficulties with wood that is too dry, stock should be seasoned to a moisture content that is optimum for the bending method and angle of curvature. For example, in hot-plate press bending, where the curvatures are relatively mild and the drying conditions severe, the optimum moisture content for the bending stock is lower than that for stock bent over forms by hand or machine. Forest Products Laboratory research shows that the optimum moisture content for bending southern oak chair-back posts in a hot-plate press is 12 to 15 percent. For chair-back rails and slats of less severe curvature than chair-back posts, the optimum moisture content for bending in a hot-plate press is probably 12 percent. For furniture parts bent over forms, stock at a moisture content of 15 to 20 percent is preferred.

Ideally, bending stock should be air-dried to the desired moisture content, generally 12 to 20 percent, and stored under controlled conditions. A temperature of 70° to 80° F. and a relative humidity of 80 percent will maintain wood at a moisture content of 15 to 16 percent. At the same temperature and a relative humidity of 65 percent, wood will remain at a moisture content of about 12 percent.

If bending stock is cut to length before drying, an end coating should be used to reduce end checking and splitting. The coating will prevent excessive absorption of moisture by the ends during the steaming process. It will also minimize the tendency to end check and split during the drying and fixing process.

MACHINING BENDING STOCK

The basic principle governing the machining of stock is that as much of the sawing, surfacing, and shaping as possible should be done on the straight piece before it is bent. In all such machining, however, several important points need to be remembered. These are: (1) Cutting stock to the minimum thickness with due allowance for deformation and shrinkage after bending; (2) accurately cut stock to length that will fit tightly in the bending apparatus; (3) surface to assure uniform thickness and remove saw marks that may induce bending failures; and (4) select rough stock with the objective of shaping the stock for bending. Due consideration should be given to direction of growth rings with respect to the plane of the bend and the ratio of width of the stock to thickness.

Since thin stock is more easily bent than thick stock, it is good practice to reduce the thickness of the bending stock as much as possible within certain limits. Stock cannot be dressed down to the thickness of the final part because it is compressed during the bending operation and shrinks during the drying and fixing process. Even air-dried stock at 20 percent will shrink appreciably after bending if used in furniture or indoor woodwork that reaches a moisture content of 6 to 8 percent in use. Boat frames, if well air-seasoned, can be cut to the approximate final thickness before bending. However, allowance should be made for final dressing to remove slight bending irregularities.

It is important to cut stock long enough to assure a tight fit in the bending apparatus so that steady and evenly distributed end pressure is exerted during the bending operation. The stock and apparatus should be long enough to allow for end trimming after bending. Stock with a long, straight leg beyond the bending zone facilitates both the designing of the straps and end blocks and the application of bending force. When several pieces are bent together to a single form, as is done with many furniture parts in hot-plate presses, all pieces should be cut to the same length. Such cutting can be conveniently and accurately done with an equalizing saw.

Careful surfacing of stock before bending has several advantages. It is easier to run straight stock than bent pieces through a planer or jointer. Removal of rough saw marks prevents many minor bending failures. Planing the stock before bending assures uniform thickness, which is especially important when stock is bent in groups. Even if surfacing is unnecessary to the final use, as in large ship frames hidden by planking and ceiling, it is advisable to surface the face that is to be next to the form. Boat and ship frames are often roughly molded before bending. Such parts of furniture as rounds and dowels are turned before being bent. It is inadvisable, however, to drill holes or cut mortises in a piece before bending.

Bending can be facilitated and breakage often reduced if the rough cutting of stock from lumber is judiciously done. Although the direction of annual growth rings with respect to plane of bend is not of primary importance in relatively mild bends, it is of practical significance in severe bends. Stock for parts requiring severe bends should be cut so that the annual rings are flatwise, or perpendicular to the plane of the bend. Such stock bends with less breakage than stock with the grain edgewise to the bending form. In bending stock with the grain flatwise to the form, it is advantageous to place the side of the piece that was closest to the center of the tree against the form, because this face is less likely to have severe surface checks.

If possible, stock should be ripped from lumber so that it is wider than it is thick. Stock that is thicker than it is wide tends to buckle laterally unless the sides are restrained during bending. When a finished part has greater thickness than width, it is frequently possible to bend wide stock and then rip it into several pieces of the proper width. Chair-back posts and rockers, which are greater in bending thickness than in width, are frequently bent in groups in hot-plate presses. The pieces support each other laterally, except for the two outside ones, which often buckle laterally unless supported.

PLASTICIZING THE STOCK

The purpose of all plasticizing treatments is to soften wood sufficiently to enable it to take the compressive deformation necessary to make the curve. Hot wood is more plastic than cold wood, and wet wood is more plastic than dry wood. A combination of heat and moisture is therefore most effective in softening wood. Treatments with hot water or steam are used commonly to prepare wood for bending. Some chemicals soften wood. However, research has not produced a satisfactory explanation of the phenomenon of plasticization, and softening methods are still based largely upon trial-and-error experience.

Softening With Steam or Hot Water

Despite considerable experimentation with various chemical treatments, plasticization with steam or hot water remains the most practical and satisfactory method of softening wood for bending purposes. Water alone softens wood somewhat, as evidenced by the fact that green wood bends more readily than dry wood. Likewise, heated wood is more readily bent than cold wood. Together, heat and moisture can produce a degree of plasticity roughly 10 times that of dry wood at normal temperatures.

In general, hardwoods are more readily softened than softwoods, and certain hardwoods more so than others. The degree of softening is one index of bending quality.

It is rarely, if ever, necessary to soften wood to its maximum degree of plasticity for bending purposes. Indeed, excessively softened wood may fail sooner than wood that is not so soft; presumably, softening weakens wood. Evidence of the effects of overplasticization is found in the results of Forest Products Laboratory tests on the steaming of white oak (table 1).

TABLE 1.—*Bending results for 1- by 2-inch specimens of Wisconsin-grown white oak,[1] steamed at different steam-gage pressures for various lengths of time and bent to 2¼-inch radius*

Steaming treatment		Bending results	
Pressure	Period	Success	Failure
P. s. i.	*Minutes*	*Number*	*Number*
0	20	37	3
0	40	38	2
0	60	37	3
17½	20	33	7
35	10	24	16
35	20	25	15

[1] Moisture content of wood before treatment was 25 percent.

The results in table 1 indicate that steaming at 35 pounds gage pressure for as short a period as 10 minutes is detrimental. Although steaming at 17½ pounds gage pressure caused more bending failures than steaming at 0 gage pressure, the number was not significantly greater from a statistical standpoint. The increase in number of failures, however, indicates that steaming at 17½ pounds gage pressure may be disadvantageous. All the specimens steamed at 0 gage pressure bent equally well, indicating that steaming at this pressure for as long as 1 hour had no harmful effect.

In other experiments, the effect of steam pressure on bending was measured by the total endwise compression steamed wood will assume before failing. Steaming at 0, 35, and 70 pounds gage pressure had an increasingly adverse effect on the amount of endwise compression that the wood could take before failing (table 2). Longer steaming periods at 35 and 70 pounds gage pressure also lowered the amount of compression the wood could take.

TABLE 2.—*Effect of steaming pressure and time on endwise-compression values of 1½- by 1½- by 3-inch specimens of Wisconsin-grown red oak[1]*

Steaming treatment		Endwise compression
Pressure	Period	
Pounds	*Minutes*	*Inch per inch*
0	20	0.353
0	40	.350
0	60	.358
35	20	.303
35	40	.309
35	60	.271
70	20	.226
70	40	.157
70	60	.105

[1] Moisture content of wood before treatment was 25 percent.

The temperature of saturated steam at atmospheric pressure, about 212° F., is generally sufficient to plasticize wood for bending. This is an advantage in several respects. The use of steam at atmospheric pressure eliminates the need for expensive high-pressure retorts. It is easier to obtain saturated steam within a closed retort or steam box when the steam is injected at low pressure. High-pressure steam becomes superheated and "dry" when released from pressure, because pressure is lost more rapidly than temperature.

The operation of a retort or steam box is simpler with steam at zero or low-gage pressure. With high-pressure steam, the steam valve must be closed and some time allowed to elapse before the door of the retort can be opened, thus delaying operations. One type of low-pressure retort is designed so that opening the door closes the steam valve and closing the door opens the valve. This arrangement speeds operations and reduces the possibility of loss of moisture from the surface of the steamed stock while the steam is off.

Steam should be injected into a retort through water standing in the bottom so that the steam within the retort will be saturated or wet. If the steam is not injected through water, the retort should be designed so that a certain amount of condensate accumulates on the bottom before running into the drain. The line leading from the boiler to the retort should contain wet or saturated steam, and a run of uninsulated pipe near the retort will also help maintain wet steam in the retort.

Treatment of wood with boiling or nearly boiling water is approximately equivalent to saturated steaming at atmospheric pressure. Boiling water is more convenient when only a portion of a stick needs to be plasticized.

It is often necessary to add moisture to wood, particularly to the surfaces, during the plasticizing process. If the stock has a moisture content of 20 to 25 percent, no additional moisture is needed, even for severe bends. At 15-percent moisture content, it is probable that no additional moisture is needed for making moderate bends. However, additional moisture is needed in the surface zones for severe bends.

To make relatively dry stock (12 percent or less) plastic enough for moderate or severe bends requires that moisture be added during the heating process. Most of the added moisture is absorbed by the surface zones. Since the concave surface must assume the maximum compressive strain, it needs optimum plasticity induced by sufficient heat and moisture. Dry wood should be both heated and moistened, but wet wood need merely be heated.

Bending stock selected from general-use lumber is probably too dry for satisfactory bending. To increase its moisture content and avoid long steaming periods, the dry stock should be soaked in water for several days. The pieces should be end coated to prevent excessive absorption by exposed end grain.

The required steaming or boiling period varies. Wet stock can be steamed for a shorter period than dry stock. Stock to be bent to a mild curvature can be steamed for a shorter period than stock to be bent to a severe curvature. Different species probably become plastic at different rates, and therefore some species may need longer steaming periods than others. As a general rule, wet stock should be steamed or boiled ½ hour per inch of thickness and dry stock 1 hour per inch of thickness.

There is some danger of oversteaming or overboiling stock, but when the steaming is done at atmospheric pressure, the danger is not great. According to table 1, stock steamed at 0 gage pressure for 20, 40, and 60 minutes produced about the same number of successful bends, 37 out of 40. Similar stock with a moisture content of 15 instead of 25 percent produced 29, 35, and 37 successful bends out of 40 after steaming at 0 gage pressure for 20, 40, and 60 minutes, respectively. For similar bending results, stock at a moisture content of 15 percent needed a steaming period of 60 minutes, while stock at 25-percent moisture content needed only 20 minutes.

In commercial bending, the plasticizing treatments should be coordinated with other bending operations. The retorts or steam boxes should have sufficient capacity to keep the bending apparatus in continuous operation. Individual retorts should be small enough to allow rapid unloading so that the surfaces of the steamed stock do not dry before it can be placed in the bending machine. Numerous small retorts also allow more flexibility in adjusting steaming periods to the thickness and character of the stock. For efficient operation, the capacities of the retorts and machines should be so balanced that the time required to load, steam, and unload the stock from the retorts is correlated with the time required to load, bend, and unload it from the bending machines. Under such conditions, the steamed stock can be placed in the machine and bent with minimum delay.

Since moisture diffuses more rapidly along than across the grain of wood, the ends of the bending stock absorb more moisture during the steaming or boiling treatment. Such pieces may end check during the drying and fixing process. Overly wet ends are also easily mutilated by the devices applying end pressure. A moisture-resistant coating applied prior to the plasticizing treatment will reduce the absorption of moisture at the ends and minimize the hazard of checking and mutilation.

Plasticizing With Chemicals

Certain chemicals, such as urea, urea-aldehyde, tannic acid, and glycerine, have been tried as wood plasticizers. It has been found that soaking wood in tannic acid solution has no important effect on plasticity.[4] Experiments at the Forest Products Laboratory using glycerine as a plasticizer failed to yield favorable results. Treatments with urea alone[5] or together with formaldehyde or dimethylolurea cause wood to become highly plastic. The effect produced by urea alone is the most marked. Urea causes wood to become thermoplastic, in which state it can be bent when hot even though

---

[4] See footnote 3, p. 4.
[5] LOUGHBROUGH, WILLIAM, KARL. PROCESS FOR PLASTICIZING LIGNOCELLULOSIC MATERIALS. U. S. Patent No. 2,298,017. Oct. 6, 1942.

the moisture content is low. Urea-treated stock does not bend so well as, and is weaker than, stock plasticized by steam.[5,6]

In limited tests at the Forest Products Laboratory, urea-treated wood was bent less successfully than steamed wood, and it developed more tensile failures during drying and fixing than steamed wood (table 3). Urea may also discolor wood and make it more hygroscopic than untreated wood. It is not definitely known how chemicals accomplish plasticization, but evidence supports the hypothesis that lignin and the less stable forms of cellulose are chemically attacked. The fact that wood can be pulped by the use of urea supports this hypothesis.

TABLE 3.—*The number of successful bends and failures in bending, and tensile failures in drying to fix the bend, in 1- by 2-inch specimens of Wisconsin-grown white oak* [1]

Chemical treatment	Heating treatment	Bending results		Tensile failures during drying
		Success	Failure	
		Number	*Number*	*Number*
None	Steam at atmospheric pressure 20 minutes.	37	3	0
Urea	Steam at atmospheric pressure 20 minutes.	25	15	4
Urea	Boiled in urea solution 20 minutes.	28	12	14

[1] Moisture content of wood before treatment was 25 percent.

BENDING

There are two broad classes of bends, those made without end pressure (free bends), and those made with end pressure. Free bending is feasible only for slight curvatures where the upset, or difference in length between the outer and inner faces of the bent piece, is less than 3 percent.

Basket rims and hoops are made by a free-bending process. Boat frames and planking are often steamed or boiled and then bent when installed on the boat by being forced into position and fastened to other framing members. Thin strips for other products are also bent without end pressure.

Of course, wood can be bent to a slight curvature without a plasticizing treatment. When it is steamed before being freely bent, however, advantage is taken of the fact that heat and moisture affect the stress-strain relation to a greater extent in compression than in tension. This makes it possible for the concave side to assume a certain amount of compressive strain before enough tensile strain is developed to cause failure in the fibers of the convex side. In bending of this nature, it may be advantageous to overplasticize

[6] HOWES, D. E., AND HOWES, D. E., JR., CHEMICAL AND MECHANICAL AIDS IN WOOD BENDING. Vt. Wood Prod. Conf. 6th Rpt.: 24–38, illus. 1946.

the wood, thereby reducing the stress-strain value in compression. Chair-back rails, slats, and pickets are often bent in a hot-plate press without applying any end pressure. Bends made by intrusion, or by forcing the stick or dowel into and through a split mold, are generally made without applying end pressure other than that obtained by friction.

Free bends are not highly permanent, even after drying and fixing, since the deformation obtained during bending is relatively slight. Consequently, it may be necessary to overbend slightly. To retain the curvature, it is usually necessary to fasten the ends of the strip together, as in a hoop or to fasten the curved piece to other members of the structure. Some free-bent members, such as chair parts, are required to retain their curvature without restraint from other members.

Bending With End Pressure

For most bending, end pressure is necessary to obtain the required compression and prevent tensile failures. End pressure can be applied in several ways, but the most common is by means of a metal strap with end fittings, such as end blocks or clamps. The metal strap is placed against the convex side of the stick to absorb tensile stress that would normally be absorbed by the wood. If the strap can be made to operate with 100-percent efficiency, no tensile stress is present in the wood, and the whole stick is subjected to compressive stress.

In actual practice, it is improbable that the metal strap often absorbs all the tensile stress. The convex face of the stick is therefore subject to a slight tensile stress and is stretched slightly in the lengthwise direction. Unless a tensile failure develops, this stretching is advantageous because it reduces slightly the amount of compressive deformation required on the concave side. To make a successful bend, the wood must be distorted without developing visible failures either in tension or in compression.

Moderate to severe bends, which require end pressure, can be divided into three general classes: (1) A simple bend in a single plane; (2) a re-entrant or S type of bend in a single plane; and (3) a compound bend in more than one plane. The different classes of bends require different methods of strapping and bending and of restraining while drying and fixing. A simple bend in a single plane may be made by hand, in a hot-plate press, or in any type of bending machine. The pieces may be bent singly, in groups, or in multiple widths to be sawed later.

The strapping of a simple bend in a single plane is relatively easy. The strap is continuous from end to end and is in contact with the convex side of the stock. The strap is either equipped with end blocks to apply pressure to the ends of the stick or clamped to the piece at the ends or at points beyond the curve. Provision may be made for adjusting the distance between the end blocks or for regulating the amount of end pressure applied. When slender pieces are bent, a means of reducing the amount of end pressure applied is necessary to avoid buckling of the unbent part during bending.

The re-entrant and S types of bends consist of several simple bends reversing in direction. Since the convex and concave sides of the stick are interchanged as the bend reverses, multiple straps are needed. For only one reversal of curvature, the straps are fastened to the form or table at the inner portions of the form and are equipped with end blocks at the outer extremities. Such bends are generally made by hand, one at a time, with the straps arranged so that they function on the convex side of the section being bent (fig. 4). The bent pieces are then clamped to the forms and dried and fixed.

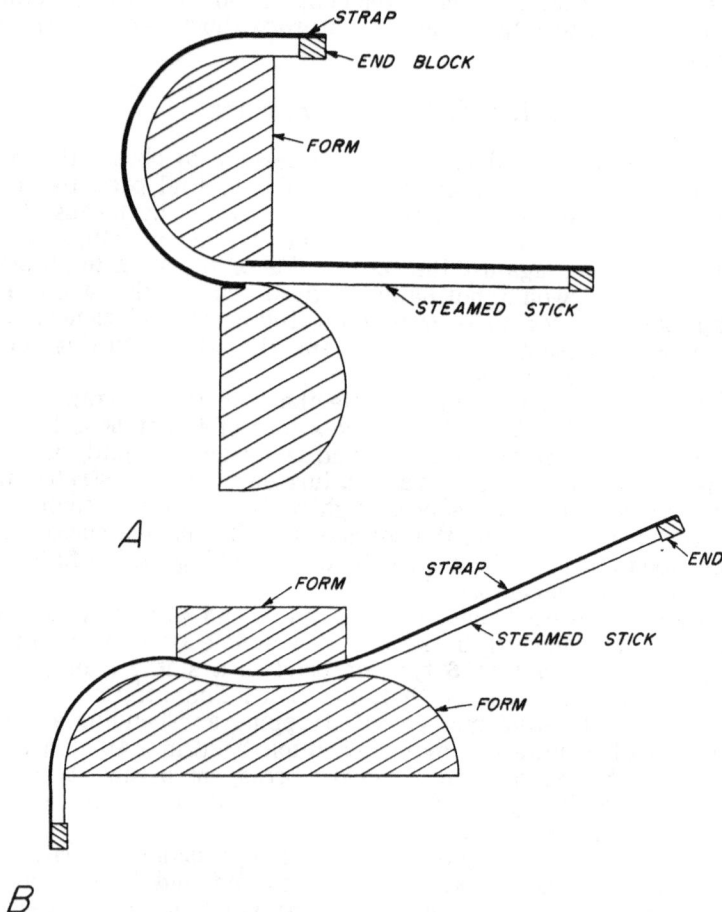

M—81679—F

FIGURE 4.—Diagrammatic sketch of bending apparatus for making (A) an S and (B) a re-entrant bend in wood.

Compound bends, which are in two or more planes, are made with a complicated system of devices for applying end pressure. Bends of this nature are made by hand by highly skilled workmen. The general scheme is to complete one part of the bend, clamp it to the form, and proceed to the next part, constantly manipulating straps and end blocks to maintain the required end pressure at all times. The combination back and leg piece of a bent wood chair is an example of this class of bend. The piece is bent to a metal form and must be dried and fixed while fastened to the form.

The actual bending technique, although subject to minor variations with the different classes of bends, has more or less general established features. After the plasticizing treatment, the stock should be placed in position for bending with as little delay as possible. In some types of equipment, the strap and end-block assembly automatically locate the piece with respect to the form. Some bending apparatus is designed so that the strap and end-block assembly hold the stock in position, preventing it from shifting on the form. In other machines, the fixed end of the stock must be clamped to the form or some device employed to hold the stock near the midpoint of the form.

The stock should fit snugly against the end blocks before bending is started. This can be accomplished by equipping the end block with a screw and bearing plate that is screwed tight or by inserting wood or metal shims between the stock and the end block. Some clearance between the end of the stock and the end block may be permissible in the early stages of bending. After bending starts, this clearance will soon disappear, and the end blocks will come into play. Generally, however, some end pressure should be applied at the very start of bending. When a number of pieces are bent together in a common strap and end-block assembly, they must be cut to a uniform length so that each will bear its share of the end pressure. Shimming individual ends is time consuming.

The bending load should be applied slowly so that the plasticized wood will be distorted more or less uniformly along its length. In some apparatus, such as hot-plate presses, the stock is prevented from backing away from the form during bending by the machine itself and the character of the imposed forces. In other types, it is necessary to hold the bent portion of the piece in contact with the form. In bending such heavy pieces as boat parts, for example, it is generally necessary to apply force to the portion already bent to keep it in contact with the form. The bending slab or table is equipped with holes for pins or dogs that are used along with wedges or screws to force the piece against the form and prevent it from backing away (fig. 5).

If end blocks are equipped with a screw and a bearing plate, it is customary to release the end pressure to some extent as bending progresses. This operation is particularly necessary when a reverse bend or buckle starts to form in the still unbent portion of the piece. The reverse bend is more likely to occur in slender pieces. Reverse bends can be eliminated by equipping the end blocks with reversed levers that extend back as far as the last point of contact between the wood and the form.

Although releasing the end pressure during bending may reduce slightly the chance of compressive failure, it invites tensile failure. Even though it has been widely recommended in the past, there is some question concerning the need for releasing end pressure as the

M—38796—F

FIGURE 5.—Bending a boat rib by pulling one end around the form and
holding the bent part against the form by means of dogs and wedges.

bending progresses, except to relieve the tendency to buckle or
form reverse bends. Measurements of end pressure during bending
tests show that this pressure increases up to a certain point and then
remains practically constant throughout the rest of the bending
process. From a theoretical point of view, the end pressure could
remain constant, diminish, or increase during a bending operation,
depending on the stress-strain relations developed in the wood.

Many devices and machines have been made for bending wood.
In some cases, the piece can be bent by hand around a form by
means of a strap and a device for applying end pressure. Some
machines consist of rollers between which the strips of wood are
passed. The diameter and position of the rollers determine the shape
of the bend. One hand-powered machine for bending several pieces at
once consists of a battery of straps and end blocks fastened to a
central steam-heated form. Each piece is placed in its strap and
bent to the form until a complete load is obtained. The pieces are
left on the form until dried and fixed. For bending heavy members,
such as ship and boat parts, a bending slab or table with the form
attached is used. One end of the piece is generally fastened, and the
free end is pulled around the form by means of tackle and a power
unit (fig. 5).

Hot-plate presses, which are widely used to bend furniture parts,
consist of hollow metal male and female forms or platens heated
by steam. The bending stock is placed between the forms, either

with or without a pan for applying end pressure, and the forms are brought against the stock by hydraulic pressure (figs. 6 and 7). A mechanical fastening is used to keep the plates in position after closing. The bent pieces are held to shape and dried between the heated plates.

M—4064—F

FIGURE 6.—A hot-plate press for bending chair parts. The pan is equipped with reverse levers and devices for applying pressure to prevent lateral buckling.

A machine commonly used to bend several heavy pieces or a larger number of small parts at the same time consists of two heavy arms connected to a heavy strap (fig. 8). As the arms move upward, the heavy steel strap pulls the stock around the form. A minor strap is generally used with this type of machine. When the bend is completed, tie rods are fastened across the ends of the minor strap to permit removal of the stock from the machine.

A machine for bending chair-seat rims or rails consists of a horizontal table with a form attached. In some machines, the form and table rotate, winding up the stock. In others, the form and table are stationary, and the stock is bent by fastening one end of the piece and carrying the other end around the form. With both types of machines, a strap and end block are used. Because the stick is slender, the end pressure is released as the bending progresses. If the stock is wound in a single plane, the ends are scarfed beforehand.

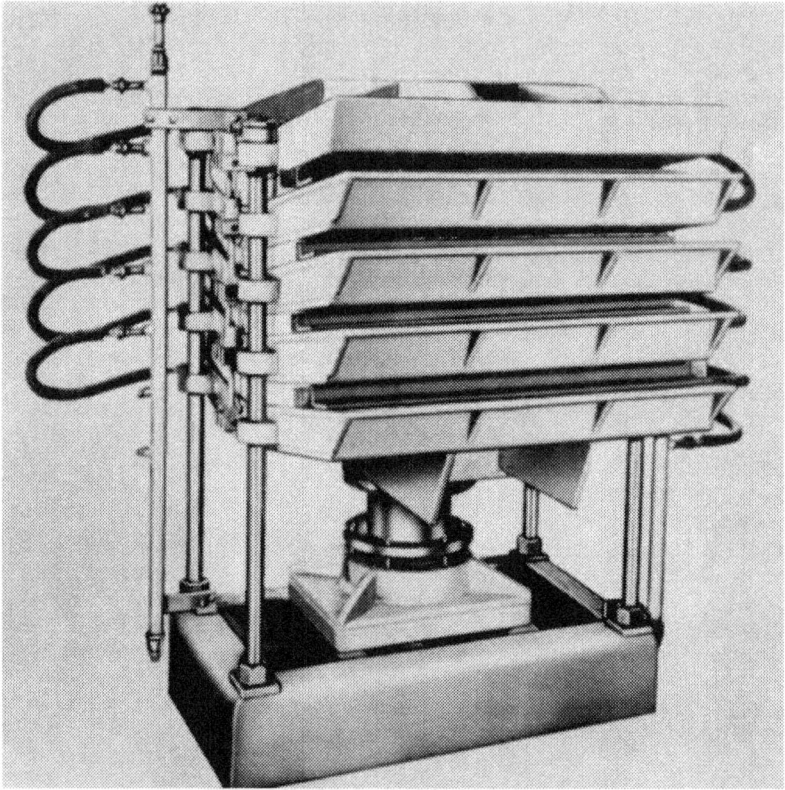

M—105254

FIGURE 7.—Another type of hot-plate press for bending chair parts.

Bending Laminated Members

The forming of curved members by laminating, which consists of bending a number of thin pieces and gluing them together, is not included in this handbook. Another form of laminating to produce curved members is analogous to the production of solid curved members. A straight piece of considerable thickness is built up by gluing together a number of thinner pieces, such as 1-inch boards. The laminated member is then bent like a solid piece. The advantages of this method are the easier and faster drying of the thin boards and the opportunity to select the best boards for the faces, particularly the concave one. However, the glue joints must be good, and the glue must be able to withstand the steaming treatment.

Applying Veneer To Improve Bending Quality

Wood of poor bending quality can be improved by gluing veneer of good bending quality to the surface that is to be concave.[7] The

---

[7] STEVENS, W. G., AND TURNER, N. A METHOD OF IMPROVING THE STEAM BENDING PROPERTIES OF CERTAIN TIMBERS. Wood 15 (3): 79–84, illus. London. 1950.

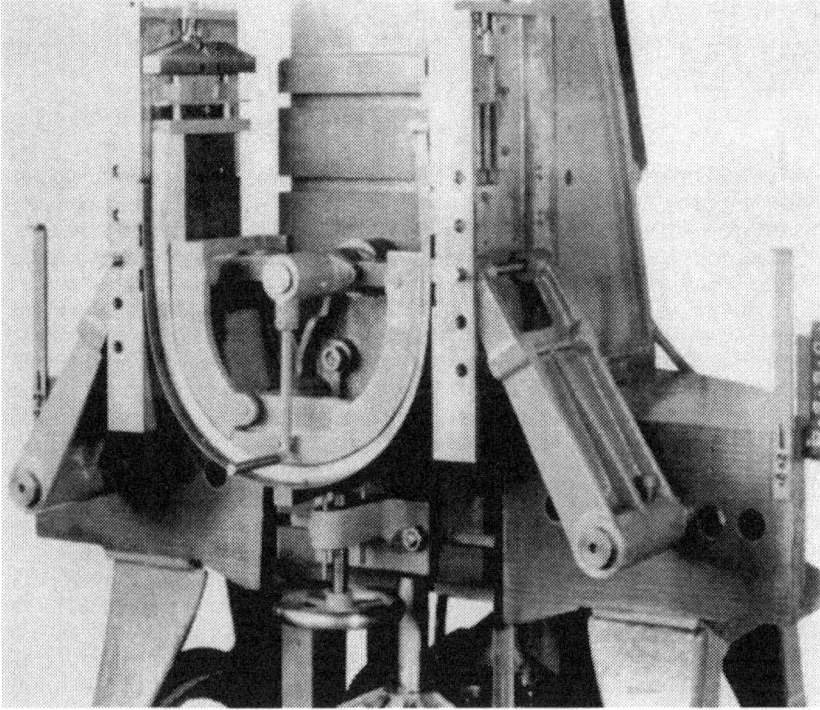

M—105257

FIGURE 8.—Rim type of bending machine with a chair part bent through an arc of 180°.

veneer assumes the maximum amount of compressive deformation and supports the inner surface of the wood.

This method was tested at the Forest Products Laboratory with ⅛-inch birch and Douglas-fir veneer glued to ½-inch sweetgum heartwood of poor bending quality. The specimens of solid sweetgum and sweetgum and veneer, all ⅝ inch thick, were at 20-percent moisture content when steamed. After steaming at atmospheric pressure for 20 minutes, they were bent under end pressure to a radius of 3½ inches. The bending results follow:

Type of specimen	Success (number)	Failure (number)
Sweetgum (solid)	6	34
Sweetgum with birch veneer:		
Grain perpendicular to the length of stick...	40	0
Grain parallel to the length of stick	30	10
Sweetgum with Douglas-fir veneer:		
Grain perpendicular to the length of stick...	23	17
Grain parallel to the length of stick	1	39

The poor bending quality of the sweetgum (6 successful bends out of 40) was vastly improved when birch veneer was glued to the compression face. The Douglas-fir veneer, however, could not assume the required deformation when the grain of the veneer was parallel to the length of the stick. Only one specimen of this group was bent successfully. Douglas-fir veneer glued with the grain perpendicular to the length of the stick improved the bending quality of the sweetgum.

Types of Bending Failures

In a successfully bent piece, the deformation, chiefly compression, is distributed nearly uniformly over the curved portion and consists of myriad minute failures. For the most part, the failures consist of folds or wrinkles in the fiber walls, with perhaps some slippage of the wood elements past each other. The wrinkling is greatest on the concave surface of the piece, and it decreases as the convex surface is approached. If the strap and end blocks have functioned perfectly, the point of zero deformation is at the convex surface. If they have not functioned perfectly, the zero point is slightly below the convex surface, which has assumed some tensile strain.

Failures occur if the plasticized wood is stressed beyond its tensile or compressive limit. In free bending, failures are nearly always in tension, because plasticized wood cannot be stretched more than 1 or 2 percent of its length. If tensile failures occur when end pressure is applied, it is because the strap and end blocks are not exerting sufficient pressure to keep the stretch of the convex side below the limit of 1 to 2 percent. This is the most common type of failure in commercial bending operations, and, as a rule, it is due to poorly designed or worn apparatus or to poor bending technique. Figure 9 shows tensile failures in a chair part bent in a hot-plate press. Such irregularities as distorted grain, cross grain, and seasoning checks may contribute to tensile failure.

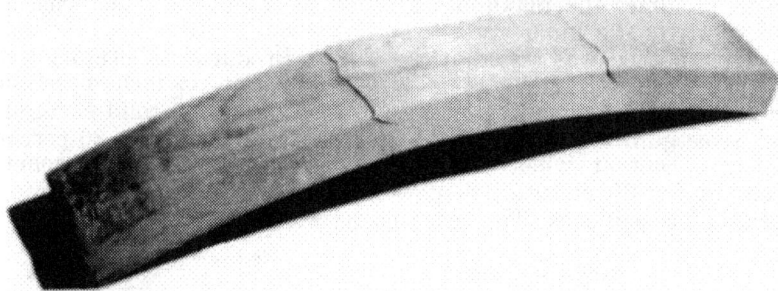

M—80832—F

FIGURE 9.—Tensile failures in a chair-back slat bent with insufficient end pressure.

Tensile failures also take the form of small slivers that break away from the convex face during bending. These are generally associated with slight cross grain. If the metal strap is as wide as the stock, it helps to prevent slivers from breaking out; if it is narrower, slivers may break out on the edges. A device for exerting pressure against the face of the stock at the point of tangency to the form is helpful in reducing slivering.

Compressive failures occur: (1) When the plasticized wood is compressed excessively; (2) when stresses are concentrated because of some defect (fig. 10); or (3) when lines of weakness encourage shear failure. The compressibility of the stock depends on the species of wood used and the plasticizing treatment. A species of poor bending quality is severely limited in the amount of compressive distortion that it can take without failing. Improper plasticizing treatments also reduce compressibility.

M—105259

FIGURE 10.—Failure in a bent boat frame. The knot could not be compressed, and the stick developed a sharp kink that caused the outer face to fail in tension.

Failure in compression may take the form of a crosswise fold or wrinkle, a longitudinal separation of fibers followed by lateral buckling, or a shear failure roughly along the longitudinal axis of the piece followed by separation of the fibers and buckling (fig. 11). The mildest form of visible compressive failure is a bulge or wrinkle extending from edge to edge on the concave side. Compressive failure may also occur in green stock. The space occupied by the excess water reduces the available void volume and causes hydrostatic pressure to develop when the wood is compressed beyond a certain point.

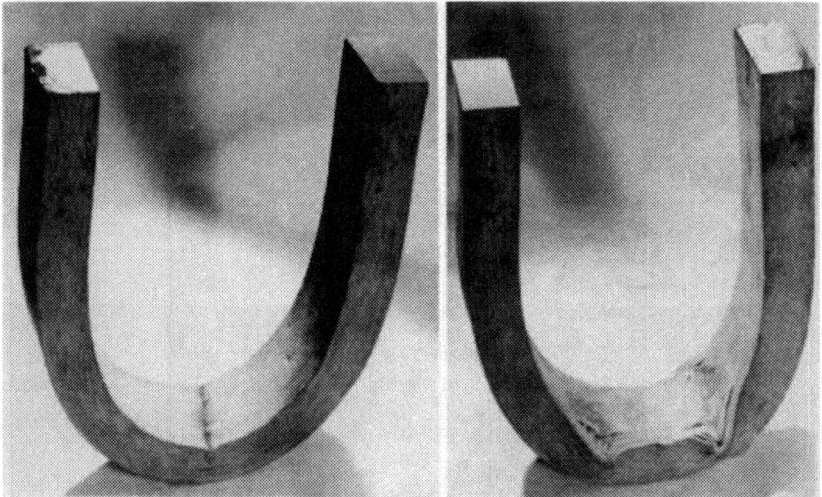

M—74784, 74785—F

FIGURE 11.—Two types of compressive failure. Cross fold or wrinkle (*left*) typical of wood of poor bending quality or improperly plasticized wood; lateral buckling (*right*) preceded by shear failure.

Compressive failures by lateral buckling in a plane perpendicular to the bend are common in pieces bent edgewise without lateral support. Such pieces are too thin or narrow to act as columns under the compressive stress, and they bend like beams. A similar effect is produced in pieces containing surface checks on the concave side (fig. 12). The surface checks set up lines of shear that, in effect, reduce the wide piece to a series of narrow strips that tend to buckle laterally, as does a single strip bent edgewise. Surface checks in combination with either spiral or diagonal grain may cause a sliver-shaped portion to shear from a corner of the piece, particularly if the checks are located near an edge or corner.

In making compound curvatures, the twisting of stock at the points where the plane of the bend changes may cause shear failure. Figure 13 shows shear failure in stock bent to form the back and leg of a chair.

Some distortion of cross section during bending is inevitable. As the piece is compressed between the strap, form, and end blocks, it tends to become thinner and wider. Unless restrained, the piece

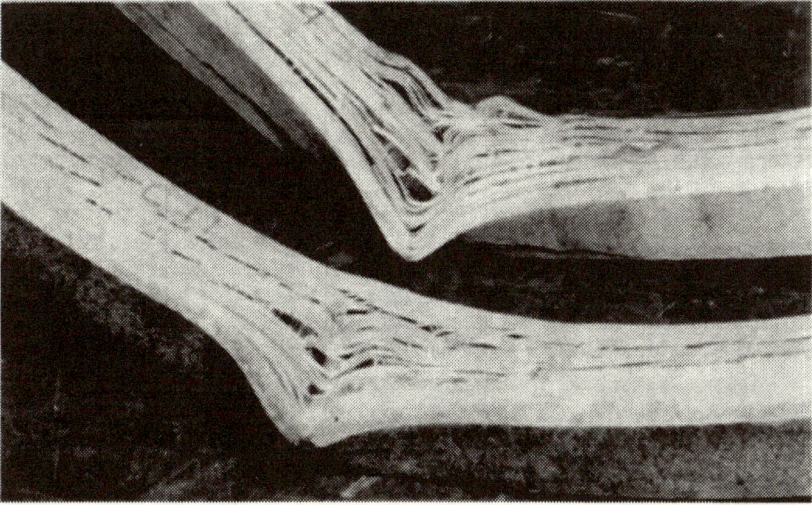

M—66664—F

FIGURE 12.—Compressive failures brought about by the presence of surface checks on concave face of stock.

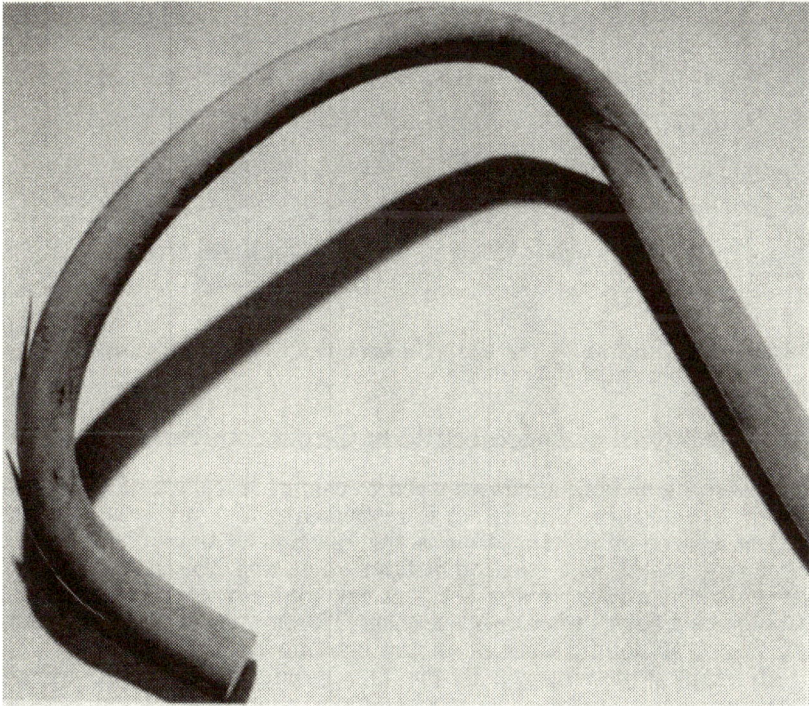

M—105178—F

FIGURE 13.—The failure in the upper right hand corner of this part is a shear failure due to twisting.

widens most on the concave side, which is in contact with the form. The increase in width is due to the fact that the plasticized wood tends to flow in the direction perpendicular to the lines of force. This increase in width probably provides some of the space needed by the compressed and folded fibers. In cutting or machining stock before bending, allowance should be made for the distortion of cross section during bending.

Mutilation of stock during bending occurs principally at the ends. Crushing or splitting of the ends is common, because pressure is not applied uniformly to the entire end surface (fig. 14). Unless end coated, the ends of a steamed piece are often plasticized more than the rest of the piece because of greater absorption and penetration of moisture during the plasticizing treatment.

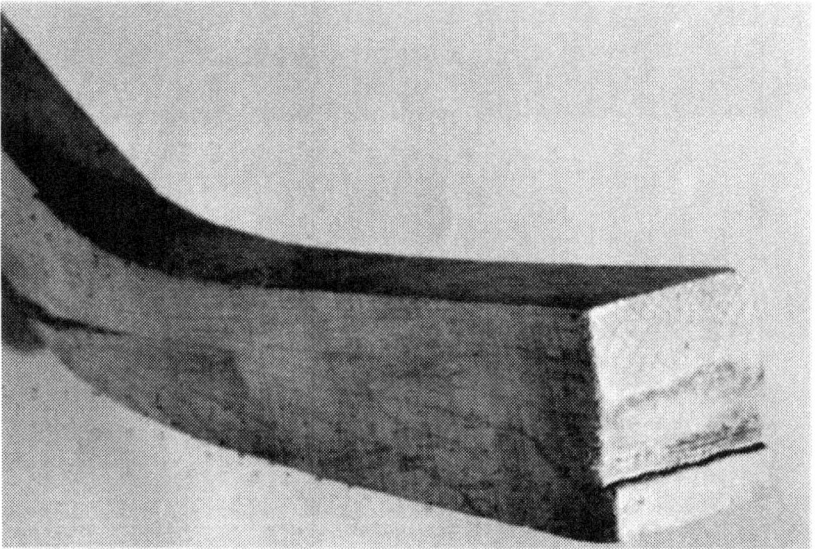

M—80833—F

FIGURE 14.—Crushing of the end of a bent chair-back post, caused by improper bearing of the end block.

Removal of Discolorations Caused in Bending

Steaming or boiling causes wood to change in color and take on a lifeless appearance. This effect is restricted to the surfaces. A more serious source of discoloration is the reaction of extractives in the wood with metal. The reaction of tannic acid and iron gives the most serious discoloration. When hot, wet oak comes in contact with iron, it becomes a dark purplish-black color. This may occur in the steaming retort or boiling tank or in the bending apparatus.

If stain is detrimental to the final product, various protective measures can be taken. The stock should be protected from drip or kept from contact with iron shelves in a retort. Although it is difficult to avoid stain while wood is soaking or boiling in steel or iron tanks, coating the inside of the tank will help. The water in the tank should be replaced at intervals to reduce the amount of iron in it. Paper or cellophane can be placed between the stock and the iron parts of bending apparatus to prevent stain. Iron or steel parts can be galvanized to eliminate most of the staining, although galvanizing is not permanent and will need to be renewed. Straps of spring brass will not stain most woods other than oak, and it only slightly.

Stain resulting from the action of tannic acid and iron can be removed by bleaching after the bent piece is dried and fixed. A hot 3-percent solution of oxalic acid applied to the stained piece will remove the stain. Afterward, the acid solution should be sponged from the piece with clear water. The process may be repeated several times if necessary.

Repair of Pieces Damaged in Bending

Failures may impair or completely destroy the utility of a bent piece. They weaken it, break the continuity of the surfaces, and impair its appearance. Tensile failures are generally more detrimental than compressive failures. Sometimes, the damage can be repaired.

If the tensile failure is in the form of a large sliver, it is occasionally feasible to force and fasten the sliver back into place, permitting the member to be used where appearance is not important. Boat parts are sometimes repaired in this way. Tensile failures in chair parts are generally cause for rejection of the piece.

Compressive failures accompanied by buckling and separation of the fibers generally make the piece useless. In chair parts with a hidden concave side, it may be possible to use a member with a rather severe compressive failure. The member can be reinforced with corner blocks. Parts with moderate compressive failures consisting of wrinkling and bulging of the fibers without much separation can often be used after the bulges are dressed off. When maximum strength is needed, defective members should not be used.

DRYING AND FIXING (SETTING) THE BEND

If a piece of wood is removed from the bending apparatus while still hot and plastic, it tends to straighten or spring back. This is a natural reaction of the wood to the release of the compressive stress imposed during the bending operation. This stress is greatest along the concave face of the piece, where some wood has failed and deformed; along the convex face it actually ceases, and a slight tensile stress may exist even during bending. As soon as the piece is out of the bending apparatus, the tensile stress intensifies, tending to pull it straight. Some compressive stress remaining in the wood along the concave face also tends to straighten the piece. However, the permanently deformed wood along the concave face prevents the piece from straightening out completely.

To counteract this tendency to spring back, the piece must be held in its bent shape until it has cooled and dried or "set." This is sometimes done by leaving it in the bending apparatus. More often, tie rods or stays are fastened to both ends, and the piece is removed from the apparatus and stored while setting. Or it is clamped to the form and removed from the bending machine.

The piece may also tend to elongate slightly with release of end pressure when removed from the bending apparatus. To counteract this tendency, a minor strap that provides some end pressure is sometimes kept on the piece until it sets, and tie rods are hooked to the ends of this strap.

With mild bends, the tendency to spring back is not entirely overcome by cooling and drying. With such bends, it is customary to overbend somewhat to compensate for the partial springback. Although cooling and drying both contribute to the restoration of stiffness and the fixing of the bent shape, drying is apparently more important.

A piece bent to varying curvature is more difficult to hold in shape. If restrained merely at the ends, it will attempt to assume a uniform curvature. It will therefore require more restraining members. The safest way to dry and fix such a piece is to keep it in the form.

A scheme that has proved helpful in retaining the curvature of bent members during drying and fixing consists of permitting the inner or concave face to dry more rapidly than the outer. This is accomplished by removing the form and retaining the metal strap or by using perforated forms. When the inner face dries first, it sets in an expanded condition along the length of the member. This set will help counteract the tendency of the bent members to close up when the entire piece reaches a low moisture content.

Bent parts are dried in various ways, depending largely on the intended use. Bent parts for boats or ships are sometimes permitted to dry on the framework of the vessel. Ship and boat builders seldom provide special drying rooms, because ship or boat framing members do not require such thorough seasoning as do many other bent parts. Pieces that are bent in hot-plate presses are dried while in the press between the steam-heated plates. The steam pressure may be 20 pounds per square inch or more, which gives a plate temperature of 260° F. or higher. In other types of bending machines, including the steam-heated form, the bent pieces are also dried in the machine.

Chair manufacturers dry bent parts in heated rooms. These rooms may be equipped with thermostats and occasionally with some means of controlling relative humidity. The temperature within these drying rooms may range from 140° to 190° F. Occasionally, bent pieces are dried in ovens at excessively high temperatures. Some bent pieces are dried in the shop by applying concentrated heat in one form or another. Many bent members are permitted to stand in the shop while drying and fixing.

Effects of Drying and Fixing

Several things happen to bent, plastic pieces as they dry. Plasticity is reduced and stiffness increased as moisture is lost. The properties of the wood becomes more like those of untreated wood, although its original strength is never completely recovered. As the wood loses moisture and plasticity and shrinks, new stresses are set up within the bent piece. The shrinkage in length and thickness causes the bent piece to attempt to take on a shorter radius of curvature. As the thickness decreases, the difference between the length of the convex side and that of the concave side calls for a general curve of shorter radius. The wrinkled and folded wood on the concave side of the bent piece develops lengthwise shrinkage in drying. This shrinkage exerts a tensile stress on the concave side that tends to increase the curvature.

If no compression member is placed between the ends or legs of the piece, the tensile stress is transmitted to the convex side and sometimes causes failures (fig. 15). Such failures can be prevented by applying some end pressure to the bent piece (fig. 16). Figure 17 shows a minor strap being fastened on the side that is to be convex. The strap will absorb some of the tensile stress during bending and during the drying and fixing process. If a minor strap or a main strap with end fittings is left on the piece during drying, any tensile stress in the convex side is taken up by the strap.

M—66500—F

FIGURE 15.—Tensile failures that occurred in a bent boat frame during drying and fixing.

Pieces dried in a hot-plate press with good bending pans are under end pressure. When no compression member is provided between the ends of the bent pieces, their curvature may be increased so that they are no longer suitable for the intended use. Overdrying increases the hazards of tensile failure and distortion of curvature. A drying room with controlled temperature and relative humidity reduces this hazard. Likewise, if the bent pieces are suitably restrained by devices that act both in tension and compression, distortions of curvature are not likely to occur.

Drying also causes shrinkage in the width of the piece, which sets up stresses similar to ordinary drying stresses. The greater the loss of moisture during the drying and fixing processes, the greater is the drying stress. The hazards of surface and end checking that were present in the original seasoning are present to a

M—66373—F

FIGURE 16.—Bent boat member prepared for drying and fixing, showing minor strap, tie rods, and wood stays.

M—105252

FIGURE 17.—Spiking a minor strap to the outer face of a piece that is to be bent to form a boat rib.

lesser degree in the drying of bent stock. The surface of wood seasoned to a moisture content of 15 to 20 percent is generally set in compression. In such a state, the wood is able to withstand exposure to severe drying conditions without developing fresh checks. The steaming or boiling treatment given wood before it is bent may, however, relieve this surface set and enable the surface to go into tension in the lateral direction when the bent piece is dried. Under such conditions, surface checks may develop in the drying and fixing process.

Stock steamed and bent at a high moisture content is more likely to develop surface and end checks during the drying process than stock steamed and bent at lower moisture content. In a furniture factory, oak and beech parts steamed and bent at a relatively high moisture content surface checked during drying. The drying room was maintained at a temperature of 170° F., with no control of relative humidity. In the case of oak chair-back posts bent in a hot-plate press and dried between plates heated by steam at 20 pounds gage pressure, the percentage of pieces rejected because of checks increased rapidly with increase in the moisture content of the bending stock. Pieces dried in hot-plate presses are highly susceptible to end checking, particularly when the ends have absorbed considerable moisture during steaming or boiling. End coatings reduce end checking.

Conditions for Drying Bent Stock

In a commercial bending operation, a variety of bent members of different species and thicknesses may be produced, perhaps with variations in moisture content. Since it is impractical to have numerous drying rooms in order to obtain optimum drying and setting conditions, all of the bent pieces are placed in 1 or 2 rooms. The temperature and relative humidity of the drying room should therefore be suitable for all of the bent stock. Temperatures of 140° to 160° F., with no moisture added to the air, are generally satisfactory unless the members happen to be particularly large, of a species difficult to season, or at a high moisture content.

The length of time that bent members need to remain in the drying room depends on the adequacy of air circulation, the thickness and moisture content of the stock, and the desired final moisture content. Bent pieces in furniture and chair factories generally are left in the drying rooms for 24 to 72 hours. A common practice is to leave them until the restraining device across the ends becomes slack. The pieces will no longer spring back when the restraining device is removed.

The time required to reach the moisture content needed to fix the bend is usually much shorter than that required to reach a moisture content suitable for service. The amount of time or drying necessary to fix a bend varies with the species of wood and the type of bend. Oak steamed and bent in the green condition becomes set even before it dries to the fiber-saturation point (about 30-percent moisture content). If a bent member is dried to a moisture content suitable for service, the bend is certain to be fixed.

Behavior of Bent Members

It is generally considered that the curvature of a bent member is permanent after it has been dried and fixed. Such is not the case, because changes in moisture content set up stresses that change the curvature. The various zones through the thickness of the piece differ in extent of deformation and in the amount of longitudinal shrinkage and swelling they will undergo with change in moisture content. In addition, shrinking and swelling in thickness tend to alter the curvature of the piece. The wrinkled and folded fibers on the concave side, which have been considerably compressed in bending, shrink and swell appreciably in the lengthwise direction. At the same time, the convex side undergoes negligible lengthwise shrinking and swelling.

The effect of the shrinking or swelling in length and thickness is cumulative. With shrinkage, the curvature is increased. With swelling, the curvature is reduced as the piece tends to straighten. Red oak specimens, 1 inch thick, steamed at a moisture content of 25 percent and bent to a radius of 2⅝ inches were dried to a moisture content of 8 percent. After the distance between the legs was measured, the bent specimens were brought to a moisture content of 21 percent. At this moisture content, the legs were more than twice as far apart as at 8 percent. When redried to 8 percent, the legs were still 60 percent farther apart than originally.

Specimens of the same material steamed at 15-percent moisture content and bent to a radius of 2⅜ inches behaved in the same manner but with even greater changes in curvature. When the moisture content of these specimens was raised from 8 to 12 percent, the distance between the legs was increased by 20 percent. When the moisture content was increased to 22 percent, the distance between the legs was 2½ times that at 8 percent. When redried to 8-percent moisture content, the legs were more than twice as far apart as they were originally when at the same moisture content.

Although neither set of these specimens maintained its curvature under changes in moisture content, the stock steamed and bent at the higher moisture content changed less in curvature with later fluctuations in moisture content. It is evident that a bent piece of wood cannot be expected to retain its curvature unless it is held at a constant moisture content or is firmly fastened to other members of the structure.

APPENDIX

Mechanics of Bending

Success in the bending of wood is dependent on the prevention by properly regulated end pressure of tensile stresses that tend to stretch the wood. This is usually done by using bending straps and end fittings. The importance of this requirement is made clear by analysis of the forces that come into action during the bending operation.[8] Figure 1, p. 2, represents a stick in the process of being bent. The stick of thickness or depth h is partially bent, and the last point of contact with the form is O. Therefore, the bending is complete to the left of O. The strap of thickness t is securely attached to the end fitting m. The stick bears through the bearing plate against a pivot at the inner end of fitting m. The line of action of P is taken at right angles to the end of the stick. If the stick and strap are assumed to be cut along a plane through O and the center of curvature of the form, the action of the portion of the strap and stick to the left of this plane can be represented by:

T, the tension in the strap acting at the center of its thickness and perpendicular to the cutting plane;

C, the summation of the stresses in the stick perpendicular to the cutting plane; and

P' the compression force at the end of the stick.

Equating external and internal moments about the intersection of the line of action of C with the cutting plane gives

$$P(X + L) = Ta \qquad (1)$$

where a is the distance between the lines of action of T and C. Equating moments about O' (the intersection of the centerline of the strap with the plane of the outer face of the bearing plate) to zero gives

$$PX = P'b \qquad (2)$$

where b is the distance from the line of action of P' to the centerline of the strap.

P is eliminated by dividing equation (1) by equation (2), and the following equation results:

$$X = \frac{L}{\dfrac{Ta}{P'b} - 1} \qquad (3)$$

---

[8] The analysis of bending stresses that is presented here was developed by T. R. C. Wilson, former member of the Forest Products Laboratory staff.

P' must be equal to the tension in the strap at O'. The tension at this point is equal to T (the tension at O) except for the friction force between the strap and the projecting portion of the stick. The friction force depends on the coefficient of friction and on the pressure of the strap on the stick. This pressure depends on the angle through which the projecting portion of the stick is bent. Since the angle ordinarily is small, the friction is small, and the tension at O' is approximately equal to T. Hence $P' = T$ very nearly, and without great error, equation (3) can be rewritten as

$$\frac{X}{L} = \frac{1}{\dfrac{a}{b} - 1} = \frac{b}{a - b} \qquad (4)$$

There is to be very little stress[9] or deformation at the convex face of the piece, and the shortening of the stick at point O will vary from zero near the face next the strap to a maximum at the face next the form. This distribution of the shortening or deformation is represented by the abscissas of the small triangle shown at O in figure 1. If the stress were proportional to the deformation, then C, the resultant of these stresses, would act at a distance of $\frac{2}{3}\,h$ from the inner side of the strap. Since the stick is strained beyond the elastic limit, however, stress is not proportional to deformation and the distance is probably slightly less than $\frac{2}{3}\,h$. It is greater than $\frac{1}{2}\,h$, however, for if it was exactly $\frac{1}{2}\,h$ no bending could occur, and if less than $\frac{1}{2}\,h$ the bending would be in the opposite direction. Danger of crushing at the end of the stick will be least if P' is applied at the center of the height or thickness of the stick, that is, if

$$b = \frac{h}{2} + \frac{t}{2}$$

Equation (4) shows that the $\dfrac{X}{L}$ ratio will be the least if a is given the largest value it can have. Since it was assumed that the distance from C to the inner face of the strap cannot exceed $\frac{2}{3}\,h$, this value is

$$\frac{2h}{3} + \frac{t}{2}$$

With the substitution of these values for a and b, equation (4) becomes

$$\frac{X}{L} = \frac{3(h + t)}{h} = 3 + 3\frac{t}{h} \qquad (5)$$

This shows that, in order to maintain approximately the proper end pressure, X (the distance from O' to the point of application of the bending force P) must be at least 3 times as great as L, which is approximately the length of the projecting portion of the stick.

---

[9] Since the strap is under combined bending and tensile stress, there may be some change in the length of its inner face. Disregarding the deformation of the convex face of the stick, which this would imply, introduces no significant error in the further discussion.

X cannot be kept constant but must decrease as L decreases. These same conclusions are reached from a more complete analysis in which the friction between the strap and stick is considered. The more complete analysis also shows that X might be kept constant if the distance b could be varied[10] during the progress of bending. The force P is not subject to variation because, with X fixed, the value of P at any stage of the operation is determined by the moment required to bend the stick at the point of contact with the form.

With the reversed-lever type of apparatus (fig. 2, p. 3), the wood is prevented from stretching, and the whole cross section is forced to assume a compressive deformation by using a thick metal strap and strong end blocks and reverse levers. Assuming that: (1) The stress curve in the wood is a straight line; (2) the metal strap does not slip on the wood; and (3) the end block is in good contact with the end of the stick at the start of the bending, the position of the neutral axis is given by

$$c = \frac{t^2 E_s + 2htE_w + h^2 E_w}{2tE_s + 2hE_w} \tag{6}$$

where c is distance from neutral axis to outer surface of strap, t is thickness of strap, h is thickness of stick, E_w is modulus of elasticity of steamed wood, E_s is modulus of elasticity of strap.

The maximum tensile stress in wood is

$$\frac{1}{2} \frac{h^2 E_w - t^2 E_s E_w}{hE_w + tE_s} \; r \tag{7}$$

where r is the radius of the form.

If no tensile stress exists in the wood, the following relationship holds:

$$t = h \sqrt{\frac{E_w}{E_s}} \tag{8}$$

This is the case shown in figure 2.

The tensile stress in the strap, and therefore the thickness, will be less than that computed by formula (8), since the stress curve in the wood is not a straight line. Therefore, the value of c is too large.

When an end block equipped with a reversed lever is used, the force required to bend the stick may be applied anywhere beyond the point of tangency of the stick to the form. The total pressure applied to the end of the stick through the bearing on the end plate that is necessary to compress or deform the wood is equal to the sum of the differential stresses over the thickness of the stick. The total pressure is equal to the tension in the steel strap. Ignoring the thickness of the steel strap, the bending force required is expressed by the equation:

$$P = \frac{Ca}{S} \tag{9}$$

---

[10] The bending can be continued to the end of the stick only by making b equal to a.

where P is the bending force, S is the distance from the point of application of P to the point of tangency of the stick to the form, C is the total compressive force, a is assumed to be two-thirds of the thickness of the stick.

White oak pieces 1 by 2 inches in dimension were steamed at a moisture content of about 30 percent for 20 minutes at atmospheric pressure. They were bent to a $2\frac{1}{4}$-inch radius through an arc of 180 degrees. One of the end blocks was equipped with a hydraulic gage for measuring the total pressure developed. The pressure increased from practically zero at the start to a maximum when the bend was about one-half completed. From this point on, the pressure remained practically constant.

The mean maximum gage pressure for several hundred pieces that were bent was about 2,200 pounds, or 1,100 pounds per square inch. At the half point of the bend, the distance from the point of attachment to the point of tangency of the stick to the form was about 17 inches. Using equation (9), the bending force, if applied at right angles to the stick, was calculated as follows:

$$P = \frac{2,200 \times 0.67}{17} = 86.7 \text{ pounds}$$

Since the pressure remained approximately constant after about the half point of the bend, the bending force had to be increased as the operation progressed, because the distance from the point of attachment to the point of tangency decreased.

The total end pressure required can be calculated for any stick if the compressive stress-strain relation for the steamed wood is known. The required end pressure can be obtained from

$$P = BR \int_{\frac{r}{R} - 1}^{\frac{r+h}{R} - 1} f\ (e)\ de \qquad (10)$$

where P is total pressure (compressive), B is width of piece, R is radius of the neutral axis of the bent piece, $f\ (e)$ is the function relating stress to strain and is obtained from stress-strain curves, e is strain, r is radius of the form, h is thickness.[11]

If the compression strain at failure is known for the plasticized stick, it is not necessary to calculate the total pressure by the use of equation (10) to determine the limits of bending. Endwise-compressibility values are rough indices of bending quality. The use of the mean endwise-compressibility value gives the calculated radius to which a given species of wood, plasticized in a specified way, can be bent with an expectation that 50 percent of the bends will be successful.

Endwise-compressibility tests were performed at the Forest Products Laboratory on 2- by 2- by 3-inch specimens steamed at atmospheric pressure for 40 minutes at a wood moisture content

---

of 25 to 30 percent. The specimens were placed between the plates of a hydraulic press and compressed until failure occurred. A strain gage attached to the central 2 inches of a specimen permitted direct reading of the unit compressive strain at the point of failure. Since the measurements were made over the central 2 inches only, it is probable that the values obtained are somewhat low. The point of failure was detected by visual means and by means of a gage indicating the maximum pressure on the ram. The following tabulation gives endwise-compressibility values determined by this method:

	Average endwise-compressibility value (inch per inch)
Species:	
Red oak (Wisconsin)	0.354
Paper birch	.287
Overcup oak (southern)	.258
Winged elm	.251
White oak (Ohio and Kentucky)	.250
White oak (Wisconsin)	.246
Yellow birch	(1)
Sweetgum	.172
Black willow	.101
Eucalyptus (Chile)	.071
Coigue (Chile)	.057
Sitka spruce	[2].015

[1] Approximately equal to white oak (Wisconsin).
[2] Approximate value.

The values in this tabulation indicate that the wood of the species with values equal to or higher than those for Wisconsin-grown white oak should be suitable for making sharp bends, sweetgum and black willow for moderate bends, and eucalyptus and coigue for slight bends. The low value for Sitka spruce indicates that it is unsuited for bending.

The mean endwise-compressibility value can be used in the following formula to estimate the minimum bending radius at which successful bends can be expected in 50 percent of the pieces for stock of a given thickness:

$$r = \frac{h\,(1 - e_c)}{e_t + e_c} \qquad (11)$$

where r is the radius of the form, h is the thickness of the piece, e_c is the compressive strain, just before failure, inch per inch, considered positive, e_t is the tensile strain associated with e_c, inch per inch, considered positive. Where an efficient strap and end blocks are used, e_t becomes zero.

When pieces are actually bent to the radius calculated by the use of formula (11), it is generally found that more than 50 percent of the bends are successful. This was illustrated in the bending of several hundred specimens of white oak, the average endwise-compressibility value of which, e_c, is 0.25. By the use of formula (11), the bending radius for 1-inch thickness should have been 3 inches. The specimens were bent to a radius of 2¼ inches with 60 percent successful bends. Had a radius of 3 inches been used, the percentage of successful bends would presumably have been even greater than 60.

It is not perfectly understood just why actual bending results exceed the expected when the bending radius is calculated by formula (11). The value of e_c is probably conservative, however, because of the method used in making the endwise-compressibility test. A zone of weakness in the block used in the endwise-compressibility test will tend to lower the value of e_c, while a similar zone of weakness may have little influence on bending failure, unless it happens to be located on or near the concave face. It is also possible that some of the deformation required to accomplish bending is provided by shear deformation. No provision for shear is made in formula (11). Consequently, the actual results when bending to the calculated radius can be expected to be somewhat more favorable than the 50-50 basis. If shear enters into bending, it means that cross-sectional planes do not remain perpendicular to the long axis of the stick.

When a piece of wood is bent, the compressive strain is probably never distributed uniformly over the length of the curved part. The distribution of the strain is presumably affected by nonuniformity of the wood along the length, local minor irregularities, friction between the stick and parts of the apparatus, and the way the stick is forced and held against the form.

Specimens of Wisconsin-grown red oak, 1 inch thick, were steamed at atmospheric pressure for 40 minutes and bent through an arc of 180 degrees to a radius of $2\frac{5}{8}$ inches. Parallel lines, 1 inch apart, were marked on the edges of the specimens before they were bent. After they were bent, the distance between these lines was measured on both convex and concave sides. The lengths of the lines across the thickness of the stick were also measured. The mean measurements for 11 sticks are shown in figure 18. The zones of maximum compression fall to the right and left of the center of the arc, but from 1 to 2 inches away from the center point. The pieces were clamped to the form at the center before bending commenced. The clamp may have increased the friction between the stick, the form, and the strap in the central portion. The increased friction may have prevented the central portion from assuming the full amount of compressive strain.

The compressive strain was not restricted to the curved portion of the stick but extended for about $1\frac{1}{2}$ inches along each straight leg. Had the stick retained its full thickness of 1 inch during bending, the total amount of compressive strain needed to make the bend would have been about 3.14 inches. According to the measurements on the convex and concave sides, the total amount of compressive strain was 2.97 inches. The thickness of the curved part, however, was reduced during bending from 1 inch to an average of 0.979 inch. On the basis of the reduced thickness, the calculated amount of compressive strain required was 3.12 instead of 3.14 inches. The difference between the calculated amount of compressive strain needed and the strain was 0.15 inch, which may have been due to shear strain.

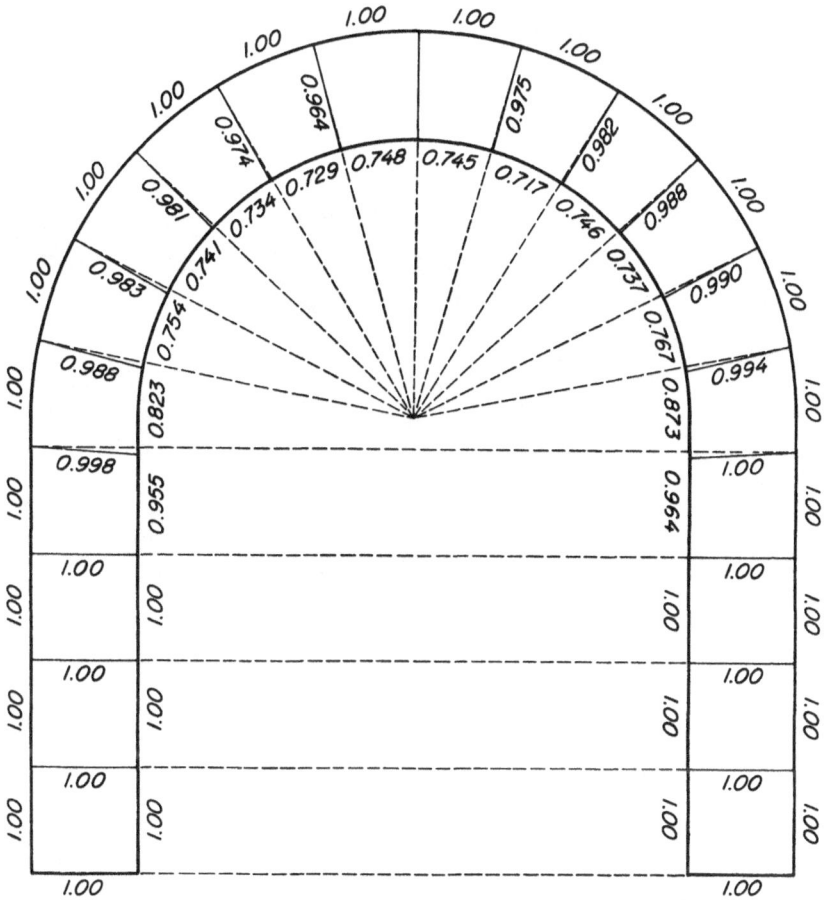

FIGURE 18.—Distribution of compressive strain in a bent stick. The mean measurements for 11 sticks are shown.

www.ingramcontent.com/pod-product-compliance
Lightning Source LLC
Chambersburg PA
CBHW022058210326
41519CB00054B/806